Table of Contents

Intro / Who This Book is For – 3

Arduino vs. "Naked" Microcontrollers – 5

A Bit of History and Quick Overview of the Arduino Uno – 8

Powering Your Arduino Uno – 14

Downloading and Installing the Arduino IDE & Tweaking the IDE Settings – 18

Getting to Know the IDE – 25

Basic Programming for the Arduino – 28

A Few Simple Projects to Get You Started – 41

Conclusion - 44

Copyright © 2018 by Custom Computer Solutions, LLC

All rights reserved. No part of this publication may be reproduced, distributed, or transmitted in any form or by any means, including photocopying, recording, or other electronic or mechanical methods, without the prior written permission of the publisher, except in the case of brief quotations embodied in critical reviews and certain other noncommercial uses permitted by copyright law. For permission requests, write to the publisher, addressed "Attention: Permissions Coordinator," at the address below.

Computer Solutions
PO Box 670863
Northfield, OH 44067
www.CircuitCrush.com

Intro & Who This Book is For

Hello and welcome to the wonderful world of Arduino, microcontrollers, programming, and electronics!

Before we dive into the meat of this book, I want to quickly introduce myself and go over a few things.

My name is Brian Jenkins. A while ago, I founded my website CircuitCrush.com because I love to learn, design, build, and hack things and I wanted to share my knowledge and projects with the world.

In 2009, I earned a degree in electrical engineering. Before that, I earned a diploma in electronics, computers, and robotics technology.

I'm also a bit of an entrepreneur. In 2007, I founded my first business (and am still in business as of this writing), a computer repair and consulting business in my area.

Enough about me.

Let's talk about who this book is for.

This is for people that are new to Arduino. If you've never touched an Arduino board before, or maybe bought one and let it collect dust because you weren't sure what to do with it, this is for you.

This is an introductory "mini course" on the Arduino Uno starting from the very basics. It's true to its title in that you can literally do it in one day and get started with Arduino.

If you've been working with Arduino for a while, this book is NOT for you, though it can serve as a quick review if you've been out of the Arduino game for a while.

One more thing to note...

Since this book is for Arduino newbies, experience with microcontrollers, programming or even electronics is not necessary, though it can be a help.

The only things you need to own are a desktop PC, laptop or Mac (which you likely already have), an Arduino Uno (here's a link to buy one if you don't already own one: http://bit.ly/GtArduinoBd) and maybe a breadboard and some small gauge hookup wire.

Your investment should be minimal.

I truly hope you enjoy this short introductory course, learn something, and have fun along the way.

-Brian

Arduino vs. "Naked" Microcontrollers: What's the Difference?

The Arduino Uno consists of the microcontroller (MCU) and various supporting components like the voltage regulator, crystal, etc.

Just like your brain is part of your whole body, the MCU is the "brains" behind the board and is part of it.

Often, trainers like the Arduino are referred to as ecosystems.

An ecosystem consists of the board and the integrated development environment or IDE.

A microcontroller is simply one of the components (the main one) on the trainer. It is not an ecosystem in and of itself, but things like supporting hardware and IDEs are readily available. They're just not wrapped up as neatly as they are in the Arduino ecosystem.

For example, when working with a naked micro, you'll need to supply the regulator, capacitors, and other components. You'll also need a compatible IDE, which you can often download for free.

For years, electronics enthusiasts did things this way and worked with stand-alone microcontrollers like the PIC or AVR.

Then came the Arduino, with the purpose of helping students and non-engineers create things, learn, and control their world. Other similar ecosystems or platforms soon followed.

While there are many pros to working with trainers, there are also some cons.

An ATmega328P, the same microcontroller that powers the Arduino Uno, lists for less than $2 in quantities of one from some vendors. If you're willing to buy a higher quantity, like ten or more, the price drops even more. Right now, the Arduino Uno (genuine model) is going for $27.95 on Amazon.

Trainers like the Arduino are great for learning and prototyping projects, but if you plan on keeping your creation for a while, a naked microcontroller might be a better choice.

Here's why:

Let's say I want to build a widget to assist me in parking my car in the "sweet spot" in my garage. The hanging tennis ball is too easy and un-elegant, so I decide to grab my Arduino and go to work.

When finished, the gadget works beautifully, and I want to keep using it.

The problem is, I have to drop another $25 on another Arduino for my next project or cannibalize my helpful parking assistant.

And if you're planning to mass-produce your project, forget it.

Suffice it to say that if you were to take your DVR or smartphone apart, you wouldn't find an Arduino inside. You would find several "naked" or stand-alone microcontrollers or microprocessors though.

The Arduino IDE is also simplified.

It's very basic as far as IDEs go. This is because to make learning simple, the designers have hidden a lot of detail and functionality behind layers of abstraction, many of which come in the form of libraries.

For example, in C programming, there usually isn't a built in digitalWrite() function (unless you write your own version).

The libraries and abstractions make doing things like reading the temperature of a sensor (an analog value) much easier for middle-schoolers and electronics newbies.

At this point, you may be thinking something like "...why even bother with the Arduino then?"

Open source electronics "trainer" boards like the Arduino and others have been an enormous boon to introducing more people to electronics and science.

This is because they make working with microcontrollers and programming easier by way of abstraction.

In the past, using a microcontroller in your projects meant working directly with the micro and supporting hardware, programming in C, and a host of other things. For those new to electronics, this could be a barrier to entry.

The Arduino and its IDE are great tools to get you started with embedded systems and programming. And, as we mentioned before, the Arduino rocks when it comes to prototyping ideas and projects.

Not only that, the boards are very versatile and can handle most common things microcontrollers can do.

Then there's the fact that the software to program it is free and can be used on Windows, Mac, or even Linux.

The hardware and the software are both open source, so others can take the designs and improve them without worrying about licensing fees.

There are many SHIELDS that are available which easily attach to the board adding functionality.

A Bit of History and Quick Overview of the Arduino Uno

A Quick History Lesson

Made in Italy, the Arduino showed up on the scene in 2005. Since then, there have been various incarnations of the board including the Duemilanove (2009 in Italian), the Diecimila (which means 10,000 in Italian to celebrate the making of the 10,000th one), the Mega 2560, and more.

Buying Your Arduino Uno

There are several types of Arduino boards that you can buy:

- Genuine Arduino boards
- Counterfeit boards
- Clone boards
- Derivative boards

Genuine Arduino boards are exactly what you'd think: the real deal made by Arduino in Italy. This is the type of board I suggest you start with. If you don't already own one, get it on Amazon here: (http://bit.ly/GtArduinoBd).

There are two types of genuine Arduino boards.

One has a surface mounted microcontroller. The other has a DIP (dual in-line package) version of the ATmega328, which is the microcontroller that powers the Uno. This is the one I suggest you get. If you make a mistake and fry the MCU, it's easier and cheaper to pop the old MCU out and replace it with a new one rather than trying your hand at surface mount soldering or buying a whole new Uno board.

As a beginner, I'd suggest staying away from Arduino clones and derivatives until you become more acquainted with Arduino and electronics in general. And you definitely don't want a counterfeit board.

As a precaution, and to avoid making a 1-day course a 1-week course due to shipping times, I'd also suggest buying one or two extra ATmega328s (DIP version, of course) in case you accidentally kill the Uno's MCU. You can buy a 4-pack with the bootloader preinstalled here (http://bit.ly/GtExtAVR).

Arduino Uno Overview

Refer to figure 1, which shows the Arduino Uno.

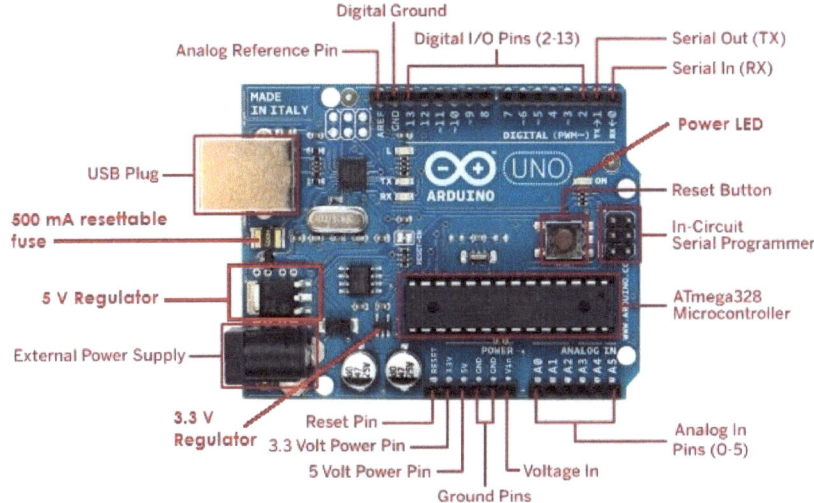

FIGURE 1: THE ARDUINO UNO BOARD

The rectangular thing in the lower right is the brains of the Uno, the ATmega328.

The power jack on the lower left is a 2.1mm center positive barrel connector. Above that, we can see a USB Type B jack for connecting the Arduino to a PC.

A series of 28 female pin headers allow you to connect other things to the Arduino and are at the top and bottom. These headers are separated into three groups. The digital pins are at the top. On the bottom left we have the power pins, and the analog pins dwell on the lower right (notice the small gap between them). Out of the 28 pins, 20 are for I/O.

The six analog pins can also serve as general purpose digital I/O. Out of the 14-total digital I/O pins, six can be used to generate PWM (pulse width modulation) signals.

The Arduino Uno also supports basic communications standards like TTL serial, SPI, I^{2}C, and 1-wire.

Two of the inputs support hardware interrupts. They trigger on either a LOW, a rising edge or falling edge, or a change in value via software.

The oval-shaped silver object in the middle left is the 16 MHz crystal.

To the right of the power jack are two 47 µF capacitors.

Directly above the jack lies a 5 V low dropout voltage regulator. Directly above the capacitors closest to the microcontroller we can see a 3.3 V low dropout regulator. It lies near two surface mount capacitors.

A Closer Look
An Arduino Uno board measures 2 1/8" x 2 ¾".

The operating voltage of the Arduino Uno is 5 V which can come from either a USB cable and your PC or some external source.

The recommended input voltage range is 7 – 12 V. You should try to keep that number closer to 7 V if you can as the small regulator will dissipate any excess energy as heat and can become very hot. We'll go into more detail on this issue shortly.

The Arduino Uno contains two regulators. The 5 V regulator can theoretically provide up to 800 mA and the 3.3 V version can source 50 mA.

We'll talk more about powering your Arduino in a minute.

Aside from the microcontroller, the Uno has a few other main points of interest.

There is an integrated USB-to-serial communications chip for downloading programs or "sketches" from your PC and for serial communications back to the PC for debugging and monitoring.

This chip is the small black box above the 16MHz crystal in figure 1.

The USB link sports a 500 mA resettable fuse to guard against any possible damage to your PC. When you plug your Arduino into a USB port, the board takes its power from that port. In USB 2.0, the current a port is capable of sourcing is usually 500 mA. USB 3.0 ports can source anywhere from 900 mA to 3 A depending on the type of port.

The analog pins connect to an internal 10-bit analog-to-digital converter (ADC). All the I/O pins can function as digital outputs and can sink/source up to 40 mA.

Next to the analog pins lie the power pins which provide access to both the regulated and unregulated power supplies.

Arduino Pin-Out and Microcontroller

The ATmega328 chip on the Arduino isn't empty, it comes pre-loaded with a small bootloader for use with the Arduino integrated development environment (IDE).

One nice thing about the Arduino is that you can use it to program other ATmega328 chips if you were to build a project with the stand-alone microcontroller.

Feast your eyes upon figure 2, which shows a pin-out diagram of the ATmega328.

One of the first things to note is that the ATmega328 pin numbers are different from the Arduino Uno pin numbers. ***This is important!***

The parenthetical labels on the outside (red writing) are any alternative uses for the pin, if any are available. The writing in red shows the ATmega328-to-Arduino pin mapping.

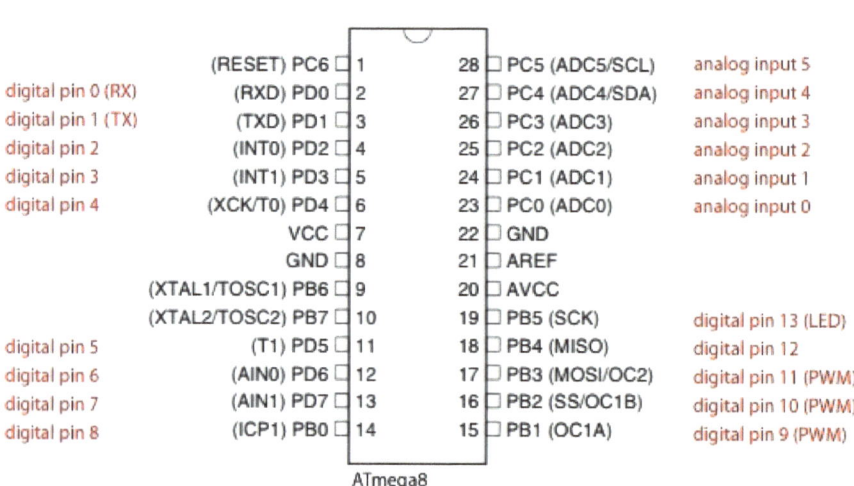

FIGURE 2: ATMEGA328 PIN-OUT AND ARDUINO I/O PIN MAPPING

Figure 3 shows a simplified block diagram of the ATmega328 microcontroller.

The I/O block is the 20 analog and digital pins. The six analog pins go to the ADC, which has a resolution of 4.9 mV because there are 1024 steps (remember, the ADC is 10-bits and that $2^{10} = 1024$) and when we divide 5 V by 1024, we get 4.9 mV per step.

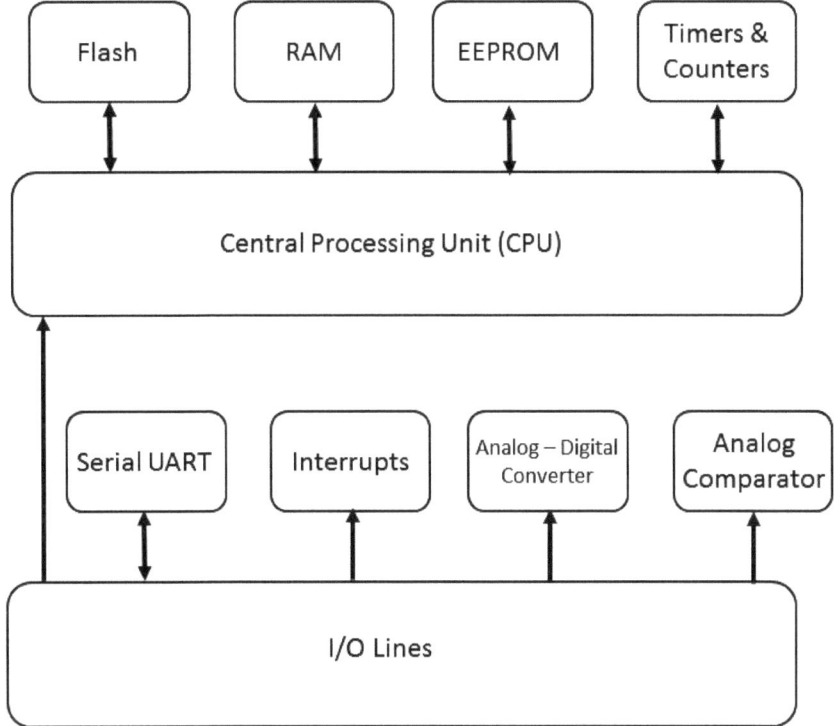

FIGURE 3: ATMEGA328 SIMPLIFIED BLOCK DIAGRAM

The two external interrupts that the chip supports map to Arduino digital pins D2 and D3.

There was a time when the analog comparator was not accessible through the Arduino IDE.

Personally, I have not tried to use this feature myself, but some quick searching seemed to suggest that there are now ways to use the comparator with the Arduino IDE.

The analog comparator will trigger an interrupt when voltage on one input equals or exceeds the voltage on another input. This could come in handy for certain projects. For example, say you want a fan to come on when the air reaches a certain temperature.

Powering Your Arduino Uno

The Arduino Uno's built-in power supply is one of the least appreciated, but most important parts of the board.

In fact, if your Arduino-based creation is giving you weird or spurious errors and you can't figure out why, there's a good chance the power supply may be the culprit.

Powering the Arduino: A Quick Overview

Figure 4 shows the power supply section of the Arduino Uno board.

As we can see, the input power can come from a few different sources:

- The DC jack (center positive)
- USB connector
- Vin header socket
- 5 V header socket

Figure 4: the power supply portion of the Arduino board.

If you plug in both a DC power adapter (like a wall wart) and a USB cable, the board automatically draws power from the source with the higher voltage. There is a small FET (field effect transistor) to the left of the crystal that isolates the USB power from the DC jack to prevent it from back feeding into your PC's USB port.

Also, there is a series diode that prevents reverse polarity from the DC jack. Inadvertently reversing the polarity can destroy the board.

The Arduino headers include three power outputs:

- Vin raw DC
- 5 V regulated
- 3.3 V regulated

The Vin socket connects to the 5 V regulator which powers the 5 V header socket and the rest of the board.

A word of caution is in order here. Vin is NOT protected by the series diode we mentioned earlier (it connects downstream of it). This means swapping ground (GND) and Vin will ruin the board.

Powering the Arduino: More Detail

You'll notice two capacitors near the DC jack in figure 4.

The one on the right stores energy for power arriving through the DC jack.

The specifications on the Uno say that you can use up to 20 V to power the board. The problem is these capacitors are only rated for 16 V. As you may have already discovered, electrolytic capacitors can fail rather catastrophically when polarity is reversed or when pushed beyond their voltage limit.

Low quality wall warts often have unregulated outputs that can exceed the ratings on the case — especially at lower currents, so be careful here.

Just above the DC power jack lies the Arduino's 5 V regulator. This device has a dropout voltage of around 1 V – 1.2 V depending on the current load. This requires a DC input of at least 6.5 V to ensure 5 V comes out, even though the Arduino specs on the voltage limits say 6 V (on the low end) is ok. Note, however, that the recommended voltage range starts at 7 V.

The power related specs (from the Arduino website) for the Uno are shown in figure 5.

Microcontroller	ATmega328P
Operating Voltage	5V
Input Voltage (recommended)	7-12V
Input Voltage (limit)	6-20V
Digital I/O Pins	14 (of which 6 provide PWM output)
PWM Digital I/O Pins	6
Analog Input Pins	6
DC Current per I/O Pin	20 mA
DC Current for 3.3V Pin	50 mA

Figure 5: Arduino Uno power specs.

The Arduino specs don't say a lot about it, but if you use more than 12 V to power the device the regulator could overheat, damaging the board.

The voltage regulator does have its own internal over-temperature protection. This will shut it down when the temperature goes above 350° F (175° C).

Once the temperature drops, the regulator will automatically resume operation. This means that powering the board improperly can not only damage it, but cause strange errors and failures as the ATMega328P microcontroller restarts.

For most of us, when we put a finger on an object of about 150° F (65° C), we will only be able to keep our finger on it for a few seconds before the pain sets in. This is a good rule of thumb for hobbyist electronics in general: if you can't touch it for more than a few seconds, it's too hot.

The Arduino, being a low-cost open-source platform aimed at hobbyists, is not meant to stand up to temperatures much greater than that for very long.

For a 12 V input, the maximum regulator current (using the rule above) is only about 70 mA.

How did I come up with that number?

I won't bore you here, but if you read the data sheet for the regulator and do a bit of math (making a few rough assumptions) you should get something close.

With a 12 V supply, the max current (using our rule) isn't much. The microcontroller itself draws about 10 mA at 16 MHz, the USB interface takes another 15 mA, and the rest of circuitry on the board (like the LEDs etc.) eat up another 10 mA or so. Not much room for anything else here.

The take-away: use a power supply less than 12 V.

For example, using the same thinking that gave us a limit of 70 mA for a 12 V power supply, we can raise that to about 250 mA if we use a 7 V power supply. Don't have a 7 V supply? Grab the closest thing to it.

This type of precaution not only protects the 5 V regulator, it also protects the 3.3 V regulator.

Downloading and Installing the Arduino IDE & Tweaking the IDE Settings

To get started, point your browser at https://www.arduino.cc/. Click the *software* tab near the top, as shown in figure 6.

Figure 6: Arduino home page.

Scroll towards the center of the page and you should see something like figure 7.

Download the Arduino IDE

Figure 7: downloading the IDE.

Note that before you can install the IDE, you'll need the free Java runtime environment installed on your PC, which you already likely have.

As of this writing, the current version of the Arduino IDE is version 1.8.5.

Websites and software versions change, so don't freak out if you come here and the site looks a bit different or the IDE version is different. The process to download and install the IDE should be similar to what you see here.

I've drawn a black box around the area on the right where you can click the link to download the IDE for your particular operating system. If using Windows, I suggest using the Windows Installer version.

As of this writing, once you click the link to install, you'll be given a chance to donate to the Arduino project. This is your call. If you just want to download the IDE, click the *just download* button.
For me, the download took about 30 seconds on my laptop, but your speed may vary depending on your Internet connection and computer.

I won't insult your intelligence by walking you through a simple software install, but I do recommend that you keep the default settings and keep clicking next. Eventually, it will ask you to install some drivers via a window that will pop up (at least it will pop up on Windows systems). Click *install.* Note that a similar window may pop up several times. Just keep clicking *install.*
When it's done, close the installer.

There should now be a shortcut for opening the IDE on your desktop.

Arduino IDE Tweaks & Settings

Once installed, open the IDE and go to *file* then *preferences*. When you do, you'll be greeted by a window like the one below.

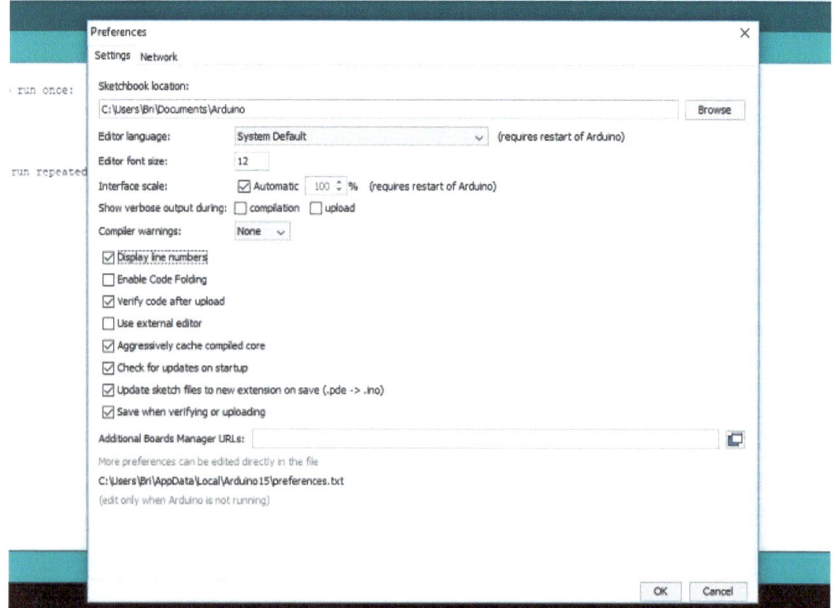

Figure 8: IDE preferences.

20

All the boxes you see checked are checked by default EXCEPT *display line numbers.* I'd check that box as it will make reading code and finding errors easier. I recommend leaving the other boxes that were checked by default alone.

At the very top, we can see the sketch book location. Sketches are the programs you write for Arduino. By default, they're stored in a directory similar to one in figure 8. You can change this to another location if you'd like by clicking the *browse* button next to it.

Next, we see the editor language dropdown. I'll be using English, but you can pick your preferred language.

Below that, you can change the font size, which you may or may not want to do depending on your preferences and vision. When done, click *OK*.

Now that we got that out of the way, let's talk about the IDE itself and connecting your Arduino.

Connecting Your Arduino Uno

When you first connect any Arduino board to your PC after installing the IDE you'll need to tell the IDE what board you're using and what port it's on.

To do this, connect your Arduino, then go to *tools,* and hover over *board.* A list of boards, as seen in figure 9 will show up. Pick the Uno, as shown on the next page.

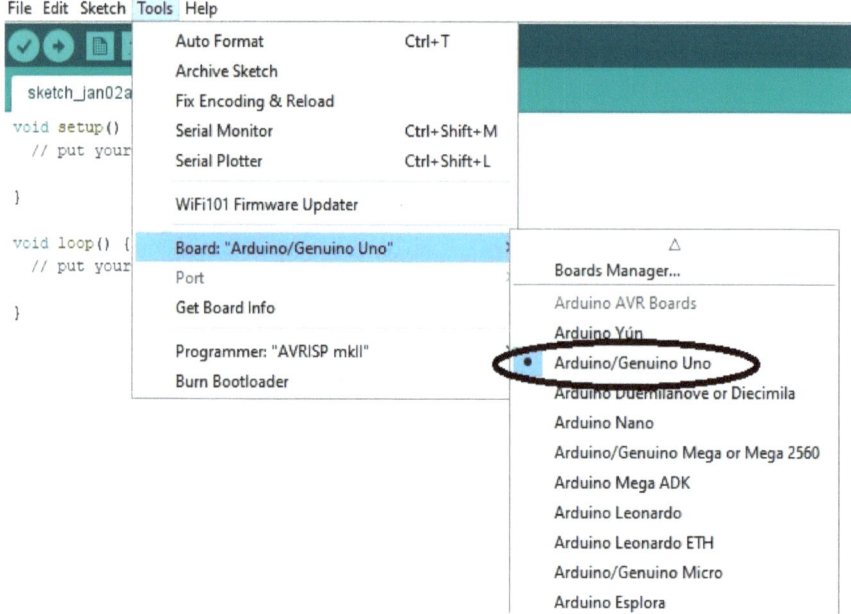

Figure 9: how to tell the IDE what board you're using

Next, we need to select our serial port. To make this easy, first disconnect the Arduino from the PC.

Now, go to *port* which is the menu item directly underneath *board* in figure 9.

Note that in figure 9 the port selection is greyed out because there was no Arduino (or anything else) connected to the laptop when I took the screenshot.

If this is the case, congratulations – this is going to be really easy. Plug your Arduino back into the PC and you should see one (and only one) port as shown in figure 10. This is the port your Arduino uses. Refer to figure 10 below.

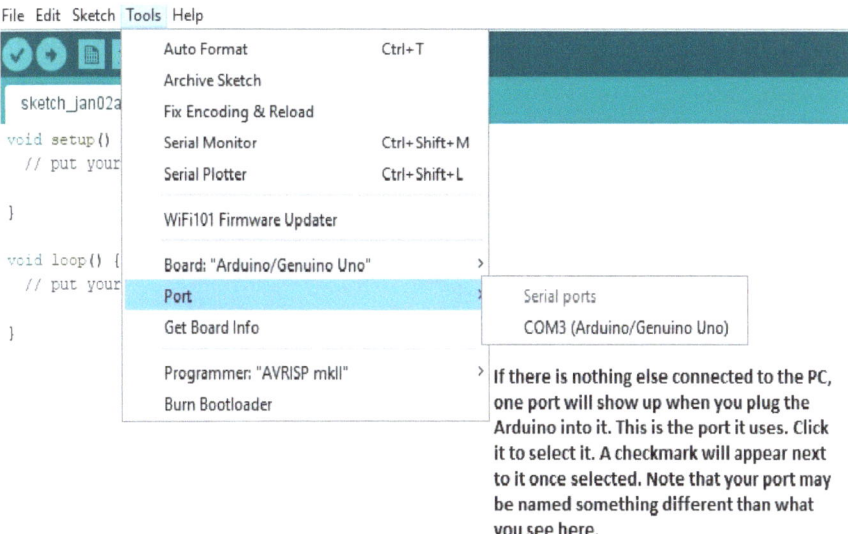

Figure 10: selecting your port if the Arduino is the ONLY device connected to the PC

If *port* is not greyed out with the Uno unplugged, you have other things connected to your PC such as a printer, external drive, etc.

Fear not.

Click on *port* and you should see a list populate. Grab your phone and take a picture of the screen or write down the ports that show on a piece of paper.

Now, plug the Uno back into the PC and go to the *port* list again. There should be a new port added to the list. Look at your picture or list to see which port wasn't there before. The one that isn't on your list or in your picture is the one your Arduino uses. Select it. In either scenario, once selected a check mark, as shown in figure 11, appears next to the appropriate port.

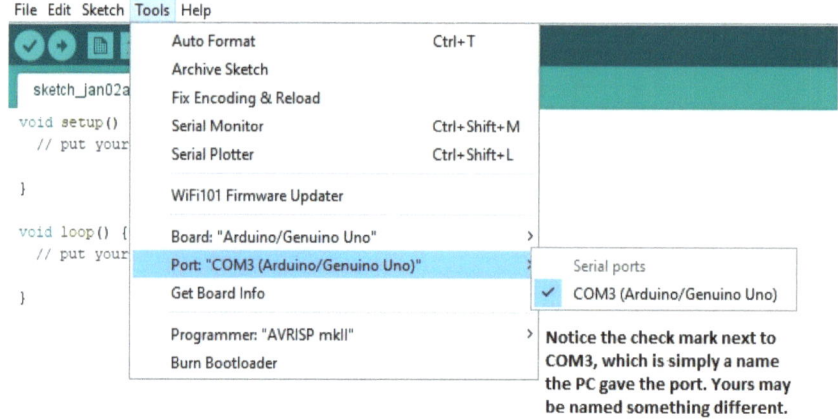

Figure 11: the correct port has been selected

You do not have to do this every time you reopen the IDE or connect the Uno; the settings should stick.

Now that we got that out of the way, let's go into a little more detail about the IDE.

Getting to know the IDE

Refer to figure 12 below for this discussion.

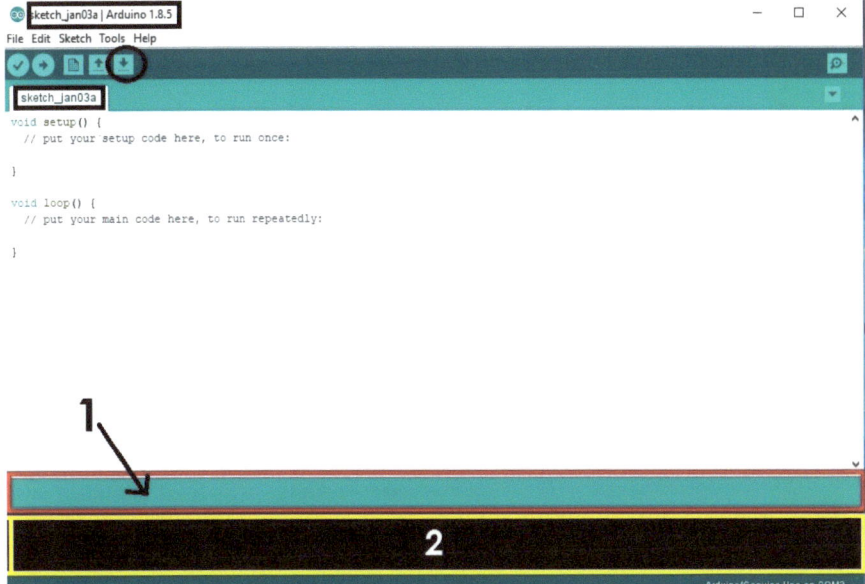

Figure 12: a quick overview of the Arduino IDE

An Arduino program is called a sketch. The black rectangles in figure 12 show the name of the sketch (in 2 different places) which by default is just the word sketch with a date next to it. The top rectangle also shows the version of the IDE.

You can change the name of the sketch at any time by going to *file* and then renaming it before saving the sketch. You can also save sketches at any time by clicking the *save* button which has a black circle around it in figure 12. When coding it's a good idea to save often.

Starting from the left of the save button, we have the *open* button, then to the left of that lies the *new* button which is used to start a new sketch.

Next, we have the *upload* button, which resembles a right-pointing arrow and is used to transfer your sketch to the Arduino. Note that two LEDs on your Uno should blink (the TX and RX LEDs; TX stands for transmit and RX means receive) when this is taking place. This is a good visual indicator. The two LEDs are shown in figure 13.

Finally, the *verify* button, which resembles a check mark, compiles the sketch. The verify button comes in handy for quickly finding errors in your code.

The IDE has three main areas: the editor, message bar, and console. Also, notice the port name and board in the bottom right. The editor is simply the white area in figure 12 where you'd type your code.

Figure 13: TX and RX LEDs

The message bar is enclosed by a red rectangle and labeled 1 in figure 12. This will give you some information about what you've recently done in the program. For example, when I upload a sketch the message "Done uploading" would appear here when finished.

The console is enclosed by a yellow rectangle and labeled 2. This is where you'll get more detail about your program and current operation and any errors. In figure 14, I've uploaded some very basic code to the board. Notice the message in the message bar. Also, the console gives me information about how much space my program takes up with other pertinent info.

Figure 14: the console and message bar

In figure 15, I've introduced an error into the code on purpose. Notice the console now.

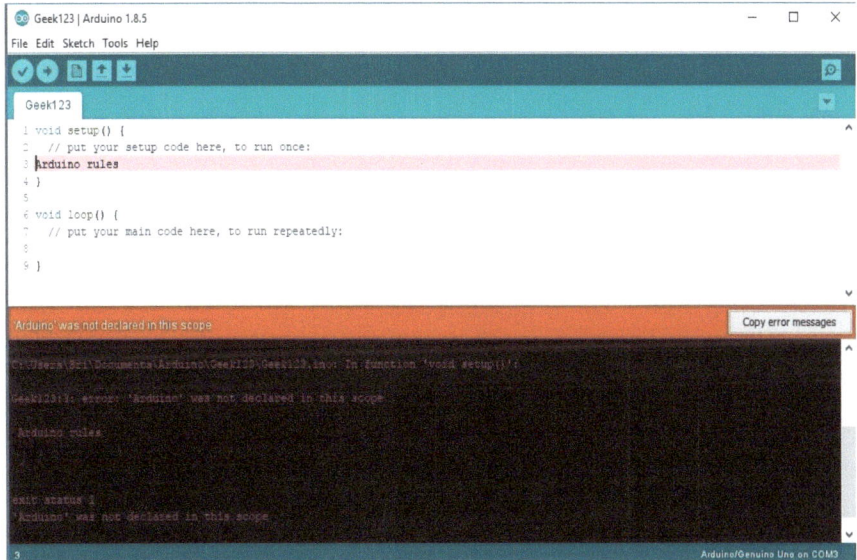

Figure 15: console with error message

There are a few things to note about figure 15.

27

First, I've changed the name of the sketch to Geek123, as we can see at the top. Second, the IDE automatically highlights the error. In this case, I inserted a comment without putting // in front of it. Third, the message bar also gives an error with the option to copy the error messages via the button on the right. Finally, more detail on the error is given in the console area.

That wraps up everything you need to know about the Arduino IDE to get started. Next, we'll do a quick overview of Arduino programming and code.

Basic Programming for the Arduino

Some of you may have no experience programming, others may be programming whizzes.

This book was meant for beginners, so I'll assume you know almost nothing about programming and cover some basics.
If you do have programming experience (especially in C) you may be able to skip this section altogether and start creating with your Arduino. Or, it may be a good review for you. Your choice.
In no way is this meant to be a complete lesson on programming, we're just covering basics to get you started.

Arduino Programming 101

Sketches are written in a language similar to C, though a sketch itself is not completely compatible with C.

There's no main() function for one thing, at least not one that's visible. It's hidden from view and added for you when you compile or verify your sketch.

If you don't know what a function is, don't worry, we'll cover that in a minute.

The sketch downloads to your Arduino via a USB cable. This process is automatic; the bootloader that resides in ATmega328 detects when a sketch is arriving.

The sketch takes residence in the 32 Kb of flash memory inside the ATmega chip. This memory can endure at least 10,000 read/write cycles, so you're unlikely to wear it out any time soon. The Arduino also sports 1 Kb of electrically erasable programmable read only memory (EEPROM) which is non-volatile. Finally, there is also 2 Kb of RAM available which is volatile.

All Arduino sketches need two parts at a minimum: the setup() and loop() functions.

These two functions take no arguments — so the parentheses are empty — but they still need to be there. We can see them in figure 12 and they are automatically inserted when we open the IDE and create a new sketch.

Just like there are rules for writing in English (or any language) there are also rules in programming. You can think of these rules as the rules of grammar for writing code.

Think of the compiler within the IDE as your English teacher from hell.

You cannot make ANY mistakes or else she'll stop reading as soon as she finds one and give you your paper back and won't even look at it again until you fix it. Then, if she sees another error after you've corrected the first the process repeats. Your paper won't get read and graded until all the grammar and punctuation are perfect. If the logic in your code is off, it may compile, but your program probably won't work as expected, giving rise to weird and hard to find errors.

Comments

Comments in programming are one of the most basic, yet useful things. When programming the Arduino, any comment starts with //. If your comment takes more than one line, each new line of it needs to have // in front of it. Below is an example of this type of comment.

//This is a comment.
//This comment is a bit longer...
//...and takes more than one line.

Long comments can be enclosed between /* and */.
/*Here is an example of a longer comment written in C this way. This can take up multiple lines in the editor, as we can see here.*/

The compiler ignores comments. They are descriptive statements meant to be read by humans, so you and others can understand your code. In other words, comments should explain what the code does. Things that are obvious may not need them, but good programming practice says to use comments generously.

Semicolons

When writing in English, you end a sentence with a period. When programming the Arduino, you end each statement with a semicolon, as shown below (exception: comments do not need semicolons).

int awesomeness = 32;

The statement above may not mean much now, but note the semicolon at the end. This tells the compiler that we are finished with a specific statement in code. If you don't use them, you'll get errors and your code won't compile.

Keywords

You may have noticed that the IDE automatically changes the color or highlights certain words like High, Low, Void, etc. These are keywords that the compiler uses and are reserved. In other words, you cannot name a variable after them. For example, the following statement would *not* work:

int High = 32;

because High is reserved for the compiler. Note that it is perfectly fine to write something like:

int High_Temp = 32;

Functions

Functions make frequently used snippets of code easy to access. The IDE has a ton of already built-in functions that are commonly used.

You can also write your own.

Let's say for example you had some code that prints the alphabet backwards every time a certain number is entered. For some reason, you find yourself using this code, which is a bit long, a lot. Instead of writing all 25 lines of it every time you need it, you can create a function that does this for you. For example, it may look something like:

Backwards_Alph(int number)
{
	//Write code that defines function here
}

Once created, all you have to do to use it is call it by typing its name:

Backwards_Alph(int number);

The first part is the name of the function and the part in parentheses is what the function takes in, in this case an integer named number. The curly braces contain the code that the function is made up of.

Earlier I said that setup() and loop() are automatically inserted when you fire up the IDE to create a new sketch. These functions have void in front of them because they return no value. Their parentheses are empty because they take in no value. Sometimes, functions do this.

All functions are followed by parentheses, even if they're empty. Inside them is where the function receives the info (it's parameters) it needs to do what it does. Some functions, like setup() and loop() take nothing, others can take multiple values separated by commas, or just one value.

A lot can be said about how functions work and how to write them. Since this is a beginner's guide, explaining functions on that level of detail would be beyond the scope of this book. I suggest starting out with the many built in functions before writing your own, unless you have programming experience.

For your convenience, this link (https://www.arduino.cc/reference/en/) should take you to the Arduino web page where all the built-in functions (and other information) are listed. Clicking on any particular function will give you more information about it.

To use (or call) these functions, simply type their name followed by (). Sometimes, you will have to put something in the (), for example:

analogWrite(pin, value);

to use this built-in function, we need a pin # and a value separated by a comma (these are referred to as the function's parameters), so I could write something like

analogWrite(5, 255);

which gives pin 5 a value of 255 (these are the function's arguments; parameters are general, like pin and value, while arguments are specific, like 5 and 255). Remember, not all functions have parameters/arguments, but they all need parentheses regardless.

Note the semicolon at the end of each statement.

For more information on this function and others, click the link referenced above.

Two Special Functions

Earlier I mentioned setup() and loop() and said that they are required in every sketch, but why is that?
What is so special about these functions? Let's talk a bit about that. The function setup() runs just once when the program starts. It takes care of things like setting up serial communications with the PC, initializing an LCD, designating certain pins as input or output, etc. Once it runs the buck gets passed to the loop() function.

A real-life analogy to this would be your morning routine. You wake up at a certain time, use the bathroom, brush your teeth, drink a cup of coffee, eat, etc. You do these things in this order once and only once at the start of each new day.

So, the job setup() performs is to execute the code inside its curly braces one time when the sketch starts.

The loop() function is somewhat self-explanatory. Once setup() runs, loop() starts, only instead of running once, loop() runs over and over again, line by line (hence the name).

A real-life analogy would be an assembly line worker. First, they get up and start their day (like the setup() function). Once they get to work, they perform the same task or series of tasks repeatedly until the day is done.

Neither of these functions return anything (that's why there is the word void in front of them), and they take no arguments, so please do not try to put anything inside the () of these functions.

Variables

Variables help us store and recall information later. They are a very important concept in programming.

A variable is the named address of a specific location in memory. Think of it as a container used to store information, like a number. The container has a name (the name of the variable) and it may contain something, like that number. The contents of a variable can change, and it often does.

Variables have a data type. Using our container analogy, this would be the type of container.

You wouldn't put a gallon of water in a paper bag, nor would you try to cram a large boulder into a gallon milk jug. The same goes with data types. When we declare a variable, we need to also let the compiler know what kind of thing we'll be putting in there, and it needs to make sense.

So, if I try to put a number like Pi into a variable with data type integer (whole number), I'm going to have problems. For example, the statement:

int Pi = 3.14;

would not work because 3.14 is not an integer. For this I'd need to do something like:

float Pi = 3.14;

In general, when declaring a variable, the following format is used:

Data Type Variable Name = something;

or if we don't want to initialize the variable with some value, we could write:

Data Type Variable Name;

Where the variable would take on some value later in the program.

ints and floats and chars, Oh My!
Integers, floats, and chars are all data types. Which one you use depends on what you need to accomplish. We'll talk about these and a few others in this section.

Let's start with **integers** or ints.

As you may know, an integer is a whole number which can be either positive or negative. This is one of the more common data types you're likely to use.

The value of an integer can range from -32,768 to +32,767. It takes of 2 bytes of data, regardless of the size of the number you use. In case you don't know, a byte is eight bits and a bit is one digital unit of data, either high or low.

Declaring an integer is really simple:

int myInteger = 800;

However, one must be careful to make sure an integer always stays within the range it's supposed to be, or weird things can happen. For example, if we add 1 to 32,767 we don't get 32,768, instead it rolls over to -32,768. This is because 32,768 is too big to store as an integer.

The lesson here is that when you create any kind of variable, you need to have some idea of what type of value it will hold and how big or small that value will get.

So, what do you do if you need to work with a number like 62,543?

There are a few options for this.

The first one is to use an **unsigned integer.** An unsigned integer can only be positive but it's range is 0 to 65,535. The example below shows how to declare an unsigned integer:

unsigned int big = 62,543;

Note that unsigned integers will also roll over to zero if 1 is added to 65,535, so be careful.

Still need an even bigger number?

No problem, just use a **long** which has a range of -2,147,483,648 to +2,147,483,647. The example below illustrated how to declare a long:

long evenBigger = 348,962,512;

A long takes of 4 bytes of data, regardless of its size. Just like the other data types, a long will roll over if the value is exceeded on either end.

You can also have an unsigned long with a range of 0 to 4,294,967,295, like the example below:

unsigned long gigantic = 4,000,000,000;

But what if we need to store a number that is not a whole number, like Pi?

That's where a **float** is used. The float data type has a huge range from -3.4028235E+38 to +3.4028235E+38. To declare a float, follow the examples below:

float Pi = 3.14159265;

float anotherFloat = 3.14E3; //this is the same as saying 3.14*10^3 which is 3,140

Floats occupy 4 bytes of space, however the bigger the float the less precise it is.

Another, smaller data type is the **byte**. Bytes take up 1 byte of space and have a range of 0 to 255. These can come in handy if you're really pressed for program space and know you value won't go out of this range. The example below shows how to declare a byte:

byte smallNumber = 16;

Next, there is the **Boolean** data type. Boolean values only take up 1 bit (not byte) of space and can be either true of false -- that's it. If you put a non-zero value into a Boolean variable, it will be considered true. For example:

boolean myBoolean = 37;

Would be true. False can be represented by a 0, the word low, or the word false.

boolean myBoolean = 0;
boolean myBoolean = Low;
boolean myBoolean = False;

//all 3 of the above statements mean the same thing and are considered false.

Similarly, the true state can be represented by the word true or the word high in addition to any non-zero number.

This data type would come in handy if a variable only needs to represent 2 states.

Finally, we have the **char**. Char stands for character and takes 1 byte of storage. Normally, you'd use a char to hold a character such as a letter or symbol. They range from -128 to +127.

There are two ways to declare a char, shown below:

char letter = 'z';
char number = 98;

Notice in the first case, we put single quotes around the letter. When declaring a char as a letter (which is actually interpreted as an ASCII value) you need the single quotes.
When putting a number in a char they are not necessary. Note that when you declare a char as a number, what's actually being stored is the letter or character whose ASCII value corresponds to that number. In the second example, 98 corresponds to a lower-case b in ASCII.

Conditional Statements & Loops

These let you do some amazing things with a little amount of code. We'll start with the if statement.

If Statement

The if statement checks for a condition and executes the proceeding statement or set of statements if the condition is true.

The syntax for an if statement is simple and goes something like this:
If (condition)
{
//do some stuff
}

For example, let's say you want to turn on a fan when a certain temperature is reached. To read the temperature, you have a thermistor (heat sensitive resistor) connected to the Arduino's ADC through one of the analog pins. Since it has a 10-bit ADC, the values will range from 0 to 1023. So, if it's cold, the value with be on the small side and if it's hot the value will be larger.

You could use an if statement like the one below to turn on the fan. This reads the ADC using the analogRead() function and then assigns that value to an integer named Temp.

```
Int Temp = analogRead(sensorPin);
If (Temp > 500)
{
//run the fan
}
```

Note that the code inside the curly braces only gets executed if the conditional statement is true. If analogRead() returns a value less than 500, the code does not execute.

An *if…else* statement is a power-up for your if statements. It allows you greater control over the flow of code than the basic if statement. This will allow you to test for more than one condition. For example:

```
If (Temp > 500)
{
//code to turn on fan goes here
}
Else if (Temp < 300)
{
//code to sound an alarm goes here
}
```

Here we turn on the fan if it's too hot as before, but now if it gets too cold we sound an alarm.

The for Loop

The for loop is one of the most useful and common pieces of code out there. It's often used with arrays, a subject that is a bit beyond the scope of basic programming but one you'll want to become familiar with. The syntax is shown below.

For (initialization; condition; incrementation) {
//code goes here...
}

At first, the for loop can be confusing and intimidating, but after a while you'll become good friends. An example from the Arduino website is given below.

```
// Dim an LED using a PWM pin
int PWMpin = 10; // LED in series with 470 ohm resistor on pin 10

void setup()
{
  // no setup needed
}

void loop()
{
  for (int i=0; i <= 255; i++){
    analogWrite(PWMpin, i);
    delay(10);
  }
}
```

Take a look at the for loop. First, we set or initialize the variable i to 0 (this part only happens once the first time we go through the loop; after the first time this statement is ignored and the other two are executed). Then comes the condition. As long as i is less than or equal to 255 we run through the loop. Integer i starts at 0, which is less than 255, so we increment i (the i++ statement increments it by one each time it passes through the loop). Then we write the value of i to PWMpin. Once i becomes greater than 255, we fall out of the loop and the code inside the loop ceases to execute.

The While Loop

The while loop is another gem you'll often find yourself using. It will loop continuously (and forever) until the expression inside the parenthesis becomes false. The syntax is below.

```
While(condition){
//do this stuff as long as condition is true...
}
```

Note that the condition is a Boolean expression and it's either true or false. The condition can be many things like some sort of less than or equal comparison, the number 1 (which means true; putting a 1 in there will make the while loop run forever), or even a function.

For example, we can write:

```
While(analogRead){
//do stuff
}
```

In this case, the loop will run as long as analogRead() returns some sort of value that's not 0.

Here's another example (also taken from the Arduino site):

```
var = 0;
while(var < 200){
```

```
// do something repetitive 200 times
var++;
}
```

The ++ operator has reared its head again and increments var every time we go through the while loop to keep count.

A Few Simple Projects to Get You Started

The Arduino IDE comes preloaded with a good assortment of sample sketches. These are to help get you started, and we'll be using two example sketches here. These were picked because neither one requires the use of any external components, and they're pretty basic.

Let's start with a version of the learning-to-program classic – the Hello World sketch (a.k.a. blink).
Figure 16 shows you how to find it.

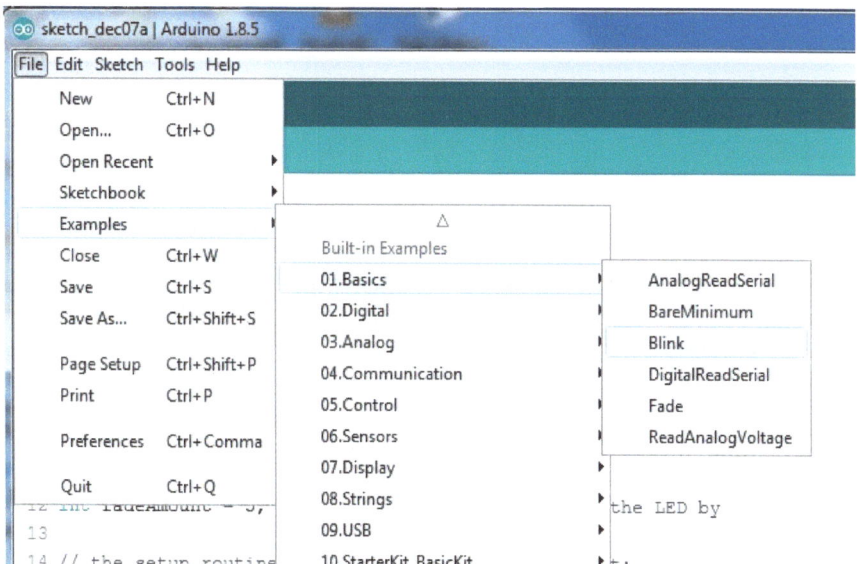

Figure 16: to get to the Blink sketch, go to File, then Examples, then Basics and click Blink.

Plug your Arduino into the PC (if you haven't already), then open the Blink sketch. Now, click *Upload* to transfer it to the Uno board. This may take a minute. The TX and RX LEDs should blink during transfer. When done the sketch should run automatically.

You should see a small LED near the top of the board blink (not the TX or RX LEDs). Looking at the code, how fast is the LED blinking? What would happen if we change the value in the delay function from 1000 to 500? What about 2000?

Now, let's look at another slightly more complicated example sketch.

Earlier, we talked a bit about ASCII. Some of you may not know a lot about it, but this sketch can help you.

Figure 17 shows you where to find this sketch.

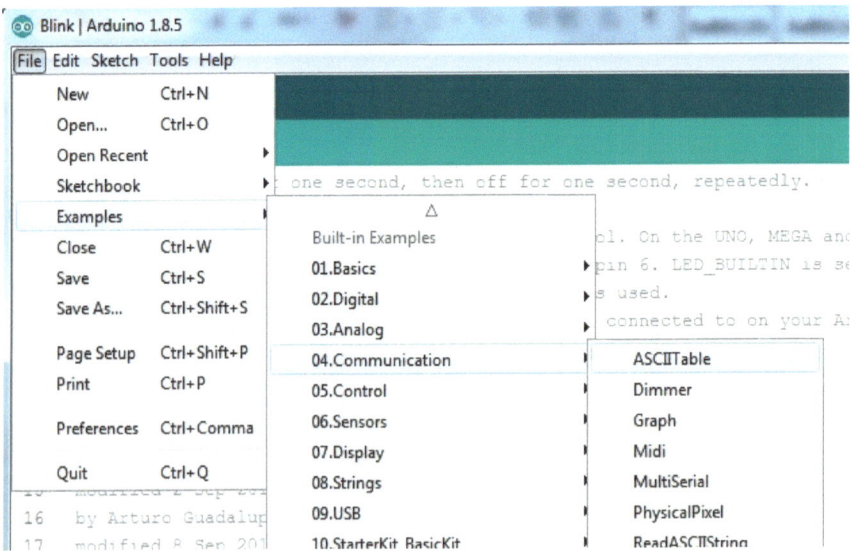

Figure 17: to get to the ASCIITable sketch, go to File, then Examples, then Communication and click ASCIITable.

Like before, upload the sketch to your board.

To see anything happen, we'll need to open the *Serial Monitor* in the Arduino IDE.

We haven't said anything about the serial monitor yet, but this sketch should give you a basic idea of what it does.

To open it, click the magnifier icon on the top right of the IDE. See figure 18.

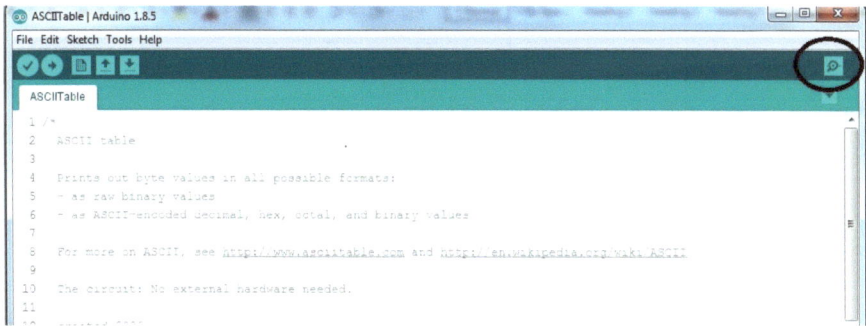

Figure 18: the serial monitor is circled in black in the upper-right of the screenshot.

Once open, the fun starts. The program only takes a second or two to finish and you should see something like figure 19.

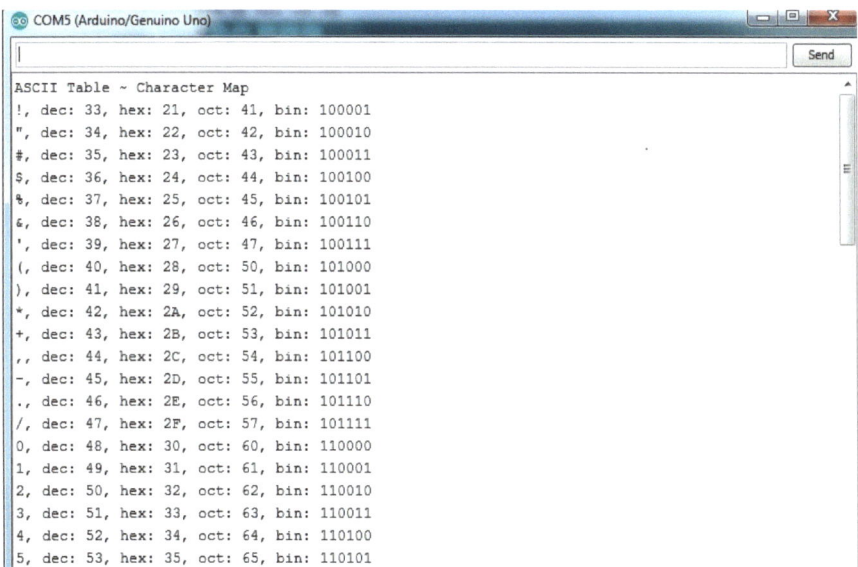

Figure 19: ASCII keyboard characters printed in decimal, hex, octal, and binary.

This program prints the ASCII characters pretty fast. How would you get it to slow down a bit (hint: think of the blink sketch)?

43

How would you get it to print the characters backwards (starting with ~ and ending with !)?
What other uses could the Serial Monitor have?

Conclusion

By now you should have what you need to get started with Arduino, programming, and the wonderful world of electronics.

While this is just a short introductory course meant to get you started, I urge you not to stop here. The Internet is full of Arduino and microcontroller based projects. There are plenty of online communities to get involved in including my **blog** CircuitCrush.com and the free **Facebook group** (https://www.facebook.com/groups/ElectronicsCrush/) where people can ask questions, share their projects, and also share news, tips, and electronics related info.

And, as you look around at things in your own life, you're bound to think of an interesting project that would somehow make your life easier, help someone, or be just plain cool.

Just a few ideas are: home automation, automotive related projects, making your own secure Cloud storage device, creating your own video game, and home security.

If you haven't already, you may want to buy a breadboard and some basic electronic components like resistors, LEDs, capacitors, etc. so you can connect your Arduino to other things. For more permanent projects, you'll need a soldering iron.

This breadboard (http://bit.ly/2BrdBd) is affordable and has lots of space.

Weller has a good name in soldering irons, and this iron (http://bit.ly/SLdrIron) is both affordable and easy to use. Though a bit more expensive, the digital version (http://bit.ly/SLdrIron2) is even easier to use.

This kit (http://bit.ly/ElecKit) is cheap and contains a good variety of parts to get you started if need be.

Until next time, keep geekin',
-Brian

www.ingramcontent.com/pod-product-compliance
Lightning Source LLC
Chambersburg PA
CBHW040249220526
45473CB00001B/426